Tackling contamination

Guidelines for business to deal with contaminated land

Centre Point, 103 New Oxford Street, London WC1A 1DU
Sponsored by ENGLISH PARTNERSHIPS

CBI Members price: £5.00 (non-members: £10.00)
ISBN 0 85201 359 0 May 1994
Design: Julian Stapleton. Typesetting: Inner Vision
Printed by Ditchling Press

Contents

How to use these guidelines 5

What is contamination? 7

Why deal with contamination? 9

Finance 9

Development 11

Property transfer 11

Insurance 12

Assessing liability 12

Working with the regulators 13

On-site risks 16

Good management and public confidence 16

How to deal with contamination 19

Initial review 19

Site investigation 20

Risk assessment 21

Remediation 24

Managing your consultants 24

Help with financing remediation 28

Helpful information 30

Main types of environmental liability 30

Funding contacts 33

Further contacts 34

ICRCL Trigger levels for some common contaminants 38

How to use these guidelines

Land contamination is an emotive subject, with overtones of ill-health and environmental hazard for the public and of large liabilities and legal fees for business. The problems are rarely this serious, but dealing with contamination and environmental liability is now a necessary part of most businesses' routine. The financial and environmental difficulties are made worse by confusion about the science, the legal framework, available technology and the validity of clean-up standards.

The CBI has been working on these issues through the political debates continuing in Whitehall and in Brussels - arguing for clear and fair law to deal with future environmental damage, and a practical and prioritised approach to dealing with the legacy of past industrial activity on the basis of the future use of the land. Much of Europe's land is contaminated - although little of it needs urgent action.

This manual has been prepared to help businesses which may be unfamiliar with the debate, but which choose to deal with the issues, either for their own benefit, or in response to an external pressure, such as a lender. It aims to enable business people to make the right decisions - not to delay tackling problems that need action, or to tackle them incompletely but equally not, as some have, to spend more than they need to achieve sensible clean-up.

The manual is structured to allow the reader to start from the trigger their own company faces - an approach from a regulator, a decision to buy or sell land, for example. In the chapter 'Why deal with contamination?' it aims to give the reader enough information to identify what steps to take next.

Decision-making in much of professional and private life is based on risk assessment of some sort, and risk assessment lies at the heart of deciding what action to take to remedy a problem. The chapter 'How to deal with contamination' outlines the approach to risk assessment of contaminated land. Beyond its simplest stages, the process is one for the relevant experts, but this guide will give a non-expert reader a feel for the language and logic, which should help in dealing with the experts if they have to be called in.

Some readers will be tackling a problem specific to their company, others may be looking at a larger area. The biggest projects to bring contaminated land back in to use are usually partnerships, often involving local and central government. The information at the back of the manual includes contacts in the public sector.

Overall, assessing the implications of contamination makes sense for individual businesses. A public commitment by business to doing so, and to dealing with problems on a sensible basis, will also help to forestall the pressures for more stringent laws which threaten to prove even more expensive for business. Environmental liability is one of the key environmental challenges facing British business in the 1990s. This manual provides first steps to meeting the challenge.

What is contamination?

Contamination of the environment occurs regularly through industrial, agricultural and domestic activity. Causes of contamination by industrial activity include:

- Spillage and leaks from pipes and storage tanks
- Disposal of waste materials on or next to a site
- Stack emissions
- Demolition of buildings with inadequate decommissioning

Contamination is often used to mean the mere presence in the environment of substances which in greater quantities would cause pollution - that is, harm to people or the environment. Contamination can of course have natural as well as man-made sources, such as the weathering of rock leading to mineralisation and from natural emissions of Radon gas. Contamination is therefore a factor, but not a sufficient condition, for pollution.

To help understand further 'What is contamination?' the terms hazard, risk, pathway and receptor, which are widely used, are defined below.

A 'hazard' exists when a substance has properties which, when particular concentrations and circumstances are present, can cause an adverse health or environmental effect.

'Risk' is the probability or likelihood that harm will occur under certain specified conditions. It is not normally possible to eliminate risk entirely. An acceptable level of contamination is determined by risk assessment.

'Pathways' are the mechanisms by which a receptor may be exposed to a hazard. Water is an important pathway which is also a receptor.

'Receptors' are the subjects which may receive or have received exposure to contaminants (e.g. people, animals, plants). Examples of hazards and receptors are discussed in the section 'On-site risks' in the next chapter.

Contamination, therefore, may pose a risk which needs to be managed. Past contamination should be investigated, the risk assessed, and where necessary remedial action taken. Guidance and advice on each of these stages is offered in 'How to deal with contamination'.

Why deal with contamination?

There are many reasons why you may wish to examine the extent of contamination on a site. This chapter outlines the main reasons and the different factors you will need to consider for each. In fact, as you assess your own priorities for action, you may find that you have more than one objective. Each is likely to demonstarte that contamination will, unless tackled, cost your business money. However, above and beyond this, land contamination is an unwelcome inheritance for future generations. We take as read the sound environmental need to tackle contamination where the levels of risk are unacceptable.

Finance

Actual or potential contamination of a site may have implications for you and your lenders. Lenders will be concerned about the impact of contamination on:

- **Your ability to fund your business:** The cost of any remediation work, compensation payments, fines, or merely the diversion of management from their core responsibilities may adversely affect cash flow and profitability.

- **The asset value of your land if used as security:** The asset value of land newly found or suspected to be contaminated will usually go down. It is even possible for the property to have a negative value when the full cost of remediation is accounted for. This is not a reason for avoiding the issue. Now that the level of awareness about contamination has been raised, uncertainty can be more damaging than having a realistic picture.

Guidance on the environmental and commercial valuation of land is offered in books 11 and 12 of the 'Manual of Valuation Guidance' published by the Royal Institution of Chartered Surveyors (RICS).

- **Their own liabilities through you:** If a lender is in direct control of a contaminated property, when realising its security, it may become legally liable for any harm caused by the contamination. Therefore lenders will wish to satisfy themselves that problems are identified and resolved. Lenders may also be concerned about the possibility of changes to the law which would have the effect of making them liable under some circumstances simply through the act of having lent money to a polluting business.

Therefore banks and other lenders are increasingly making enquiries about the environmental liabilities of prospective borrowers or existing clients whose facilities need renewal or upgrading. Issues the lender may consider include:

- The borrower's awareness of the potential consequences of its environmental liability, including present and historical uses of the land
- The extent to which the borrower is managing any risk
- The quality of the borrower's environmental management.

As some forms of contamination may migrate and affect the surrounding property, the lender may also take into account the past and present uses of land next to your own site.

DSC Foundries Ltd

DSC Foundries, a subsidiary of the Sugg Lighting Group, planned in 1993 to expand its metal processing activities, and purchase a larger site. It located a suitable disused factory on an industrial estate in Greenwich. The factory had previously been used by metal sprayers, but had gone into receivership.

During the negotiations for the purchase of the freehold, DSC Foundries' bank insisted that a full environmental site investigation was necessary before a loan could be secured. The receivers were reluctant to grant access to the factory for the site investigation to be carried out.

DSC Foundries, therefore, conducted its own initial review. It obtained plans and surveys of the site from Greenwich library, dating back to 1850 when the land was used for farming. Houses were built in 1917 and remained until 1960 when the metal spraying factory was established. DSC Foundries knew which processes were conducted on the site during the occupancy of the factory. This information was presented to the bank, via a surveyor, and was sufficient to satisfy their lending requirements. The purchase of the site is now going ahead as planned.

If you think that environmental liability is likely to be, or should be, an issue with your lender, the first step is to find out your lender's requirements as soon as possible. In the case of banks, the first approach should be to the branch or office you usually deal with. If they are uncertain, encourage them to clarify their company's policy. It is worth remembering that some lenders prefer to use their own consultants for any survey work, so check their preferences before commissioning any assessment intended to support a loan application.

Development

There is no legal obligation for a planning applicant to volunteer information on contamination on its initial application form. However, a Local Planning Authority can refuse an application for planning permission if unacceptable contamination of the land is suspected.

A Local Planning Authority has the power to insist on receiving such information as it requires to determine the application, including a full site survey. The CBI, RICS and the Council of Mortgage Lenders propose that future relevant planning applications should include a land appraisal.

If in doubt, it is sensible to try to present the planning authority with all the information it might need to make its decision. Presenting too little might lead to delays or a more expensive request.

Property transfer

The law places rights, obligations and liabilities for both vendors and purchasers of a property. The sale and purchase agreement, which is a vitally important document in any transfer, can modify the way in which these are divided.

The legal basis of all property contractions is caveat emptor - let the buyer beware. In the absence of any other agreement, and subject to the Property Misdescriptions Act, (1991) a purchaser who buys a property leaves the vendor with no residual liability to the purchaser (it does not follow that the vendor has no liabilities elsewhere). However, for example, should a purchaser ask a vendor whether a piece of land contains contamination and the vendor states that it doesn't (although the vendor knows that it does) the purchaser should be protected from liability as misrepresentation has taken place in the sale.

In any transfer involving significant contamination, the need to be sure about what liabilities are being taken on may affect the terms of the contract. It may be possible for a purchaser to obtain agreement that the vendor will be responsible for past activities which lead to liabilities after the sale. Alternatively, the vendor might reflect the purchaser's potential liability by reducing the price.

The types of provision that may be incorporated into a contract of sale include: warranties, indemnities, covenants and guarantees. Clearly, detailed drafting is a matter for lawyers. Particular attention should be given to agreement of the definition of key words, much as environment and contamination.

If the property is to be purchased on leasehold, the physical state of the site should be established in the terms of the lease by means of a baseline survey. The freeholder should be informed of changes in site use during the tenancy.

The Construction Industry Research and Information Association (CIRIA) is to publish an excellent guide to the sale and transfer of land which may be affected by contamination, which sets out the steps in much greater detail. Perhaps the key message is the need for both purchaser and vendor to reduce their future unforeseen liabilities by commissioning a joint or separate baseline survey where there is any likelihood of unacceptable contamination.

Insurance

Insurance is not intended to provide indemnity for known contamination. Its role is to offer financial protection for fortuitous damage. Insurance cover therefore cannot provide a complete solution to contamination risk, but should be one component of a total environmental risk management strategy.

There are two main types of policy:

- **Public liability policies.** Most public liability policies written after 1991 contain a specific pollution exclusion clause. The Association of British Insurers recommends exclusion of all contamination claims except those caused by "a sudden, identifiable, unintended and unexpected incident which takes place in its entirety at a specific time and during the period of insurance."

  Most public liability policies now attempt to exclude 'gradual' type pollution and limit cover to pollution claims arising from 'sudden' pollution matters. Policies should be carefully covered for exclusions and expected coverage.

  Old public liability policies without a specific pollution exclusion clause may continue to provide valid cover for pollution that manifests in due course, provided policies were written on a 'claims occurring' basis.

- **Specialist environmental impairment liability insurance.** Specialist policies have been developed to cover some of the liabilities excluded from the public liability market, especially gradual pollution. But they are generally site-specific and are tightly under-written.

Claims are usually limited to a specified sum on an aggregate and claims made basis, with a pre-insurance site investigation required. Thus the policies may be difficult to obtain and appear comparatively expensive.

Some industrial sectors may begin to look to forms of insurance or financial guarantees on a mutual basis. New legislation increasingly looks to ensure that a business operating today will have the resources to deal with tomorrow's problems.

It is important to be aware of what your insurance covers. In particular, if in doubt, keep and review old public liability policies, to establish their applicability to pollution problems that become apparent well after the event and the time at which they were written.

Assessing liability

Contamination may pose the risk of harm to humans, the environment and property. Such harm may lead to:

- **Criminal and/or civil liability under statute.** The two principal statutes which create environmental liabilities are the Environmental Protection Act (1990) and the Water Resources Act (1991). The consequences of being found liable depend on the offence and may include a fine, a duty to pay for remediation work carried out by a regulatory body, or a legal requirement to remedy the damage. For criminal offences, proceedings may be brought by any interested party, including pressure groups. No actionable loss need be demonstrated.

Directors may face personal liability and companies may be found liable for actions taken by employees, contractors, or vandals in some cases.

- **Civil liability under common law.** Common law proceedings may be brought by parties which have suffered an actionable loss. The consequences of proven liability are damages and costs which may be of any size. Many cases are settled out of court. The court's interpretation of the evidence in each case is particularly hard to predict. The recent Cambridge Water Company case, taken to the House of Lords helped to clarify the uncertainty.

Under statute and common law the liable party could be the person responsible for the contamination, the owner of the land, or the occupier. The Government is considering consolidation of various statutory powers, and the European Union may also legislate in the future. Nevertheless, some areas of responsibility are becoming clearer.

The steps you may need to take depend on the nature of the potential liability. A fuller description of the principal liabilities is given in the chapter 'Helpful Information'.

Working with the regulators

Regulators have considerable powers to deal with urgent pollution problems, and they may identify the problems through their effects on surface and groundwater as well as the land itself. The principal players are:

- National Rivers Authority (NRA) - with responsibilities for surface and groundwater. (River Purification Boards in Scotland).
- Her Majesty's Inspectorate of Pollution (HMIP) - which regulates specified industrial processes through integrated pollution control.
- Waste Regulatory Authorities (WRAs) - which are mainly county councils in England and district councils in Scotland and Wales, and have responsibilities for controlled wastes.
- Borough and District Councils - which are responsible for environmental health, air pollution control and dealing with statutory nuisances.

The regulators carry out these responsibilities in the context of their wider duties and there is some overlap. For example, a 'deposit' might be a statutory nuisance - the responsibility of the borough council; a controlled waste - the responsibility of a WRA; and cause water pollution - the responsibility of the NRA. Some overlap will be reduced under the single Environment Agency which will combine the responsibilities of HMIP, NRA and WRAs, but will not include all Local Authority pollution control powers.

The regulators have powers and responsibilities under statute to prosecute offenders and to sue for recovery of reasonable expenses (see earlier section 'Assessing liability', and 'Helpful information' on the main types of environmental liability).

In most cases, the regulators will do their best to take into account the resources and commitment of a company which is causing or has caused pollution. Therefore, if a regulator is likely to become involved, it is worth exploring the problem with them at the earliest possible time, so that no effort is spent unnecessarily and so that the regulator can reflect your company's willingness to deal with the issue in any decisions about the use of its statutory powers.

In 1987/88 Esso closed and disinvested its lubricants blending and packaging facility in Manchester. The site had been in use since the early part of the century, when construction and operating standards were less stringent, and awareness of land contamination issues much lower than today.

The consultants Dames and Moore carried out preliminary investigations to establish the extent, levels and types of contamination, and to provide data for the risk assessment process, ie. migration pathways, aquifer characterisitics, vulnerable receptors, leach tests etc.

Esso felt there would be two key elements to the success of the project:

1. satisfying the regulators (NRA) that the preferred bio-remediation techniques would be effective and have no unacceptable consequences. Clean-up criteria needed to be determined using the future water quality objectives for the adjacent river as a basis (part of the Mersey Basin improvement scheme), and taking account of the soil characteristics and groundwater regime on the site.

2. satisfying potential buyers of the land that it was fit for commercial use, and that they were not taking on a liability, by identifying all the relevant information and making it available to them.

Esso contacted the NRA at an early stage and as soon as they had obtained information on the condition of the site and identified the preferred bio-remediation options. Following discussions with the NRA a model was developed for predicting the groundwater and contamination migration for the post-remediation state using the results from on-site testing. The model also took account of the additional reduction in contamination that would occur due to continual biological activity after switching off the clean-up process. By this means it was possible to agree a cost-effective level of clean-up to be achieved by the bio-remediation processes and to avoid expenditure and effort in unnecessary reduction of contamination.

The clean-up on site involved major civil engineering works. As a first step a 200 metre long cut-off trench was installed to isolate the redundant area from the continuing operations. Then over 1300 cubic metres of contaminated soil were excavated and treated on specially prepared beds by a landfarming process designed to accelerate normal degradation of hydrocarbons. Biological clean-up of the groundwater was achieved by abstraction down-gradient, dosing with nutrients, then reinjecting through the cut-off trench. Site works lasted about 16 months.

The outcome was therefore a comprehensive remediation and monitoring exercise which:

a) had the support of both the site owner and the regulators at the outset

b) incorporated a review of progress at the end of each phase

c) avoided costly recycles

d) provided some assurance to buyers by demonstrating that the owners had discharged their Duty of Care by agreeing clean-up criteria and remediation measures with the regulators in advance.

On-site risks

Employees, construction workers, neighbours and future users of the land may be affected by contamination by:

- Inhalation of vapours or particulate matter (which may also be blown off site)
- Ingestion (eg. eating soil, particularly by children)
- Indirect ingestion (eg. eating plants grown on contaminated soil).
- Direct skin contact
- Explosions or fire from flammable gases

As an employer it is important to ensure that employees' exposure to hazards is either prevented or - where this is not reasonably practicable - adequately controlled. The most relevant statutory provisions include the general terms of the Health and Safety at Work Act (1974) and the Control of Substances Hazardous to Health Regulations (1988).

Further guidance is provided in 'Protection of workers and the general public during the development of contaminated land', produced by the Health and Safety Executive.

Good management and public confidence

Good environmental management will allow your business to minimise the risks arising from contamination and improve your dealings with all those with an interest - lenders, insurers, employees, customers, suppliers, potential purchasers and the public. It will also enable you to minimise the risk for the future, both by managing the environmental effects and by being able to demonstrate that you have taken a responsible approach towards the environmental impact of the operations of your business.

Avoiding contamination now is better, and probably much cheaper, than dealing with it later. Plant modifications and good housekeeping can help avoid contamination. Particular attention should be given to drainage systems and soakaways, unloading and transfer points, storage areas and waste disposal facilities.

You may like to consider adopting one of the developing environment management systems, such as BS7750, or the EC Eco-Management and Audit Scheme. The CBI's Environment Business Forum itself aims to provide businesses with help, advice and examples on good environmental practice and can, if you wish, also form a first step towards applying for recognition under one of the statutory schemes. Individual industrial sectors are also increasingly providing guidance for their members.

Environmental management usually involves improved record-keeping and can lead to greater public reporting by businesses of their environmental activities. This may carry with it the risk of greater short-term exposure to liability, but may also increase confidence in your business, and help to persuade government against imposing more burdensome reporting or regulatory requirements.

Tackling contamination in a thorough and professional manner will allow you to be open with appropriate information to all those with an interest in your business. This openness will remove potential misunderstandings and is as important to the effective management of this issue as the technical concerns which follow.

British Gas

British Gas owns over 900 operational sites in the UK with a total land area of approximately 2800 hectares. Many of these are former town gas works and are potentially contaminated. Although much contamination was removed at the time of plant demolition, various by-products of the gas-making process, such as spent iron oxide, used for removing sulphur and other contaminants from the gas, and tar along with other waste residues are likely to be present to some degree on any former gas-making site.

British Gas recognised that a comprehensive survey programme was required to gather more information about the potential risks caused by the presence of contamination on these sites. During 1992 all the sites have been categorised according to their potential for causing off-site pollution.

As a result, British Gas identified the most sensitive sites and conducted site investigations to establish the condition of each site at its perimeter. Several sites have been identified as having cross boundary contamination and discussions have taken place, and are continuing to take place with the NRA to determine the prioritiy sites for remediation. An environmental provision of £158 million has been made in the company accounts to pay for the remediation programme.

Tackling contamination

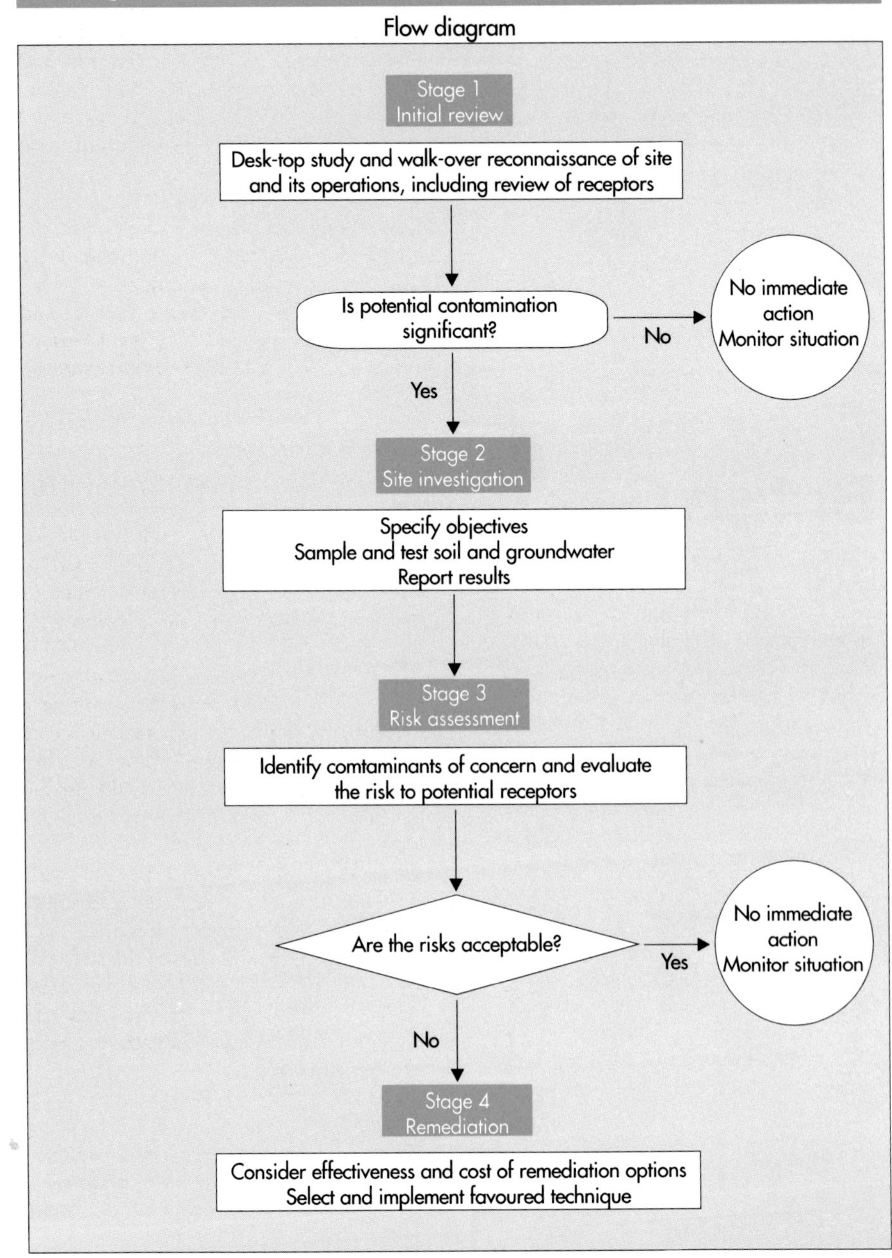

How to deal with contamination

If you have now decided that you need to look further at the possibility of taking action, you will need to adopt a strategy for an evaluation of the hazards and assessment of the risk. The flow diagram illustrates the recommended stages of investigation, risk assessment and remediation.

You should be able to do some of the initial review and decision-making yourself, or to have them done in-house. The later stages require expert advice and this guide covers some of the language and logic you will need to deal with the experts.

Initial review

The first stage of the strategy is an initial review of the site. The primary aim is to identify the nature and extent of contaminants which are likely to be present so as to determine whether to proceed further with a site investigation.

However, the presence of contaminants in itself does not necessarily constitute a hazard. The physical characteristics of the site, the potential receptors and the purpose of use of the site should also be considered.

Clear objectives should be set with differentiation made between assessment based on past/current activities and assessment for future uses of the site. The requirements will be different for a site which is to continue in industrial use from one intended for residential property redevelopment.

Method

The review should consist of a desk-top study in conjunction with walk-over reconnaissance of the site. The purpose is to establish:

- Past and present industrial operations at and adjacent to the site to identify possible contaminants and their distribution
- The layout of the site above and below ground including buildings, storage areas and services
- Potential receptors at risk of harm from contamination, both on and off the site.

Sources of information for this review include:

- Company records (past and present)
- Maps: including Ordnance Survey (OS) topographical, geological and soil surveys
- Regulators records: local authorities, water authorities and waste disposal authorities
- Meteorological and hydrological records
- Photographic records (including OS aerial photographs)
- Long serving employees

Is a site investigation necessary?

The nature of the identified contaminants, potential receptors, and the end use of the site will form the basis of the decision as to whether a site investigation is needed. Assessment of the contaminants should consider whether they are:

- Toxic
- Volatile
- Carcinogenic
- Mobile
- Persistent
- Bio-accumulative.

This information about the contaminants allows an assessment of the potential hazards and will help determine the significance of the risk each contaminant poses to humans, the environment or buildings. This is a judgemental form of risk assessment contrasting your findings with available published guidance. Since you will need to do this again if a full site investigation is needed, further guidance is included in the section on 'Risk Assessment' later in the manual.

Where a site investigation is considered necessary, the initial review can be used to design the sampling and testing procedure.

If at this stage you take the view that it is not necessary to proceed with a full site investigation, it is important to document the steps taken and justification given. If necessary, put in place a monitoring programme to ensure that the decision can be reviewed if circumstances change later on.

Site investigation

A site investigation is intended to provide a detailed evaluation of the hazards posed by contaminants and form the basis for a risk assessment. It should provide information on the extent of contamination, its mobility (a very important issue in determining the need for action), rate and direction of migration.

The investigation requires sampling and chemical analysis of the soil and groundwater (and surface water where appropriate) to establish the magnitude and distribution of site contamination. This is carried out under a sampling plan based on the information accumulated in the initial review. Sampling techniques include soil cores, trial pits, boreholes, geophysical methods and passive gas surveys. Information about the geology and hydrogeology of the site should also be collected. If you do not have the in-house expertise, you will need to seek assistance, such as that from consultants, and the section 'Managing your consultants' offers guidance on their selection and instruction.

The following summary outlines the principal stages of a site investigation:

- Examine findings of initial review
- Set objectives for site investigation
- Identify necessary safety precautions
- Design sampling programme for identification of both contamination and geology/hydrogeology
- Collect representative samples
- Analyse samples
- Present analytical data in a report

The report should summarise the sampling procedure and analytical methods used. Its purpose is to state the extent of contamination and indicate the likely direction and rate of movement. The significance of the results should be determined by an assessment of the risk. That assessment must be based on the end use of the site.

Risk assessment

Although risk assessment is ostensibly a simple process, a detailed site-specific assessment is difficult to carry out in practice, especially where it involves environmental science which evolves rapidly. The process is complicated by the lack of agreed standards against which to assess the significance of the contamination.

At each stage of the assessment, it is prudent to review whether enough data has been obtained to complete a reliable assessment of the risk from which remedial options (if any) can be derived. In only a very small number of complex cases will it be necessary to proceed to a full site specific quantitative characterisation of the risk to receptors.

The first stage of the assessment is to consider whether contaminant concentrations are of any concern at all and to eliminate from the investigation those which are not. This can be done by comparing concentrations found on site with guidance values available.

Where such guidance values are used, they must be treated with care. Several sets have been developed and in all cases their context and purpose must be understood and used appropriately.

Among guidelines commonly used in this country are:

- **The UK Interdepartmental Committee on the Redevelopment of Contaminated Land (ICRCL):** which set out threshold and action values applicable only to the redevelopment of contaminated sites. These values are for guidance and do not determine the quality of the soils or background environmental concentrations.

  If all the sample values are below the threshold triggers then the site may be considered uncontaminated for the end use. If some of the sample values exceed the action triggers then the site may require remediation for the proposed redevelopment of the site to take place. Where sample values lie between the threshold and action values, objective judgment is required.

  The ICRCL standards cover only a limited list of compounds, and take no account of effects on groundwater. The ICRCL trigger values for some common contaminants are summarised in 'Helpful information'

- **The Dutch Intervention Values:** which set soil criteria intended for all uses of land. Thus sample values below these triggers are suitable for any purpose based on background levels for unpolluted soils in the Netherlands. The triggers are not intended as standards for remediation and care should be taken when applying the values for guidance in this country, not least because they may lead to greater expense than is necessary.

Given the difficulties caused by the lack of standards, the Department of the Environment is putting considerable effort into a research programme designed to give better information about assessing environmental risks, and guidelines for remediation based on the intended use of the land. Detailed risk assessment should take account of the outcomes of this work programme. A regularly updated bibliography of published research by the DoE and other bodies is available from the Association of Environmental Consultancies.

The presence of contaminants of concern in itself does not necessarily constitute a risk. The characteristics of the contaminants, the pathways and the receptors potentially at risk should be analysed together.

Thus the assessment should consider:

- Nature of the contaminants (eg. toxicity and mobility)
- Proximity and sensitivity of receptors (e.g. rivers and groundwater)
- Site-specific criteria (eg. permeability)

This will enable the determination of the receptors' potential exposure rates to the hazard. Acceptable levels of contamination at specified receptors may be determined with reference to existing standards and guidelines (eg. statutory water quality objectives, drinking water standards).

Whilst the assessment of risk is rigorous in its analysis of the nature of contaminants and soil, the process requires the use of safety factors at certain stages to account for any uncertainties. Therefore the results will be conservative, erring on the side of safety. This should be carefully considered when making comparisons with existing guidelines.

The end point of the assessment process is a risk characterisation report, which may be presented in the following ways:

- General statements concerning the condition of the site and the potential for harm
- Comparison with published guidance (eg ICRCL)
- Computed values using numerical models
- Reference to statutory standards (eg. drinking water quality)

WIMPEY ENVIRONMENTAL

BUSINESS DECISIONS • FOR THE FUTURE

A site containing a non food retail outlet was to be conveyed from one pension fund holder to another. Since the site is located in an industrial area of Plymouth the two institutions decided to fund jointly a site investigation.

Following agreement on warranties to the funds and tenants, they commissioned Wimpey Environmental to undertake the investigation.

Wimpey undertook an initial desk based study to establish previous industrial uses of the site and potential contaminants that could arise from their processes. It examined the geology and hydrogeology to determine the sensitivity of the site and the surrounding environment to any contamination that might be present.

Following this study it devised a programme of investigation and analysis to address the potential contaminants which had been identified and confirm the nature of the geological conditions.

In order to minimise disturbance of the site, it used a driven probe system to undertake the sampling exercise. It undertook soil gas analysis to detect the presence of hydrocarbon contamination and hazardous gases, and took soil samples for visual examination and subsequent laboratory analysis of solid phase contaminants.

The results of the studies were used to formulate a risk assessment for the site, which was itself incorporated into a land quality statement. ICRCL values were used as guidance for the assessment of on-site risks. Off-site migratory pollutants were assessed on the basis of the geological and hydrological studies.

This work showed the site did not present a significant risk to the site users, surrounding areas or the groundwater environment. On the basis of the statement the conveyance of the site was successfully completed.

Remediation

The risk characterisation report will help determine whether remediation is necessary and which remediation options are appropriate for your site. There are three main categories of remediation techniques:

- Containment

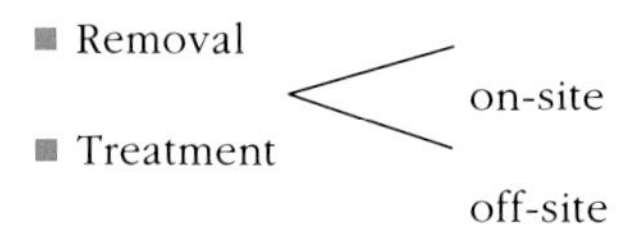

Only a few techniques can completely remove or destroy all contamination. Complete removal may not be necessary where an acceptable final level of risk, without it, has been identified by the risk characterisation report. The choice of technique by you or your advisers should therefore be clearly set against the actual/ intended use of the site. For example, containment may be the most suitable technique where the end use of the site is a car-park, but if houses are to be built removal or treatment is likely to be a more appropriate technique.

The risk assessment may indicate that remedial action is not required immediately. If this is the case a suitable monitoring programme should be set up to ensure that such a decision remains valid over time. Where a site is grossly contaminated changing its end use to a less sensitive one may prove the most appropriate action.

Where remedial action is required, factors such as local hydrogeology, soil type and the volume of material to be treated will affect the choice of technique to be adopted. You should also consider the regulatory requirements of the NRA, local authorities and Health and Safety Executive, and take care to ensure that remediation techniques adopted do not compound the original problem, for example creating new pathways to groundwater.

It is possible that no single technique will deal with the contamination cost effectively and hence a combination of techniques may be used. Exhibit One summarises the range of remediation techniques currently available. The treatment techniques can be further subdivided between those which can be carried out in-situ (on site) and those which require removal and treatment ex-situ (off site).

Depending on the complexity of the chosen technique(s) you may need to employ the services of contractors. However, responsibility for remediation rests with your company. You should make sure that all remediation work carried out is fully documented using photographs and sample testing as is appropriate.

Upon completion you should establish a monitoring programme to ensure that the hazards of contamination have been adequately dealt with.

Managing your consultants

Consultants are often required to carry out the site investigation and risk assessment. This section offers guidance on selection and management.

Selection

Unless you have already established links

Exhibit One: Summary of remediation techniques

Technique	Method	Examples	Cost (1993 order of magnitude)
Containment	surface capping	tarmacadam synthetic liner imported soil	£35-75/m2
	vertical barriers	diaphragm walls impermeable walls sheet pile walls	£60-£180/m £25/m £20/m
	horizontal barriers	jet grouting	£280/m2
Removal	off-site disposal	landfill	£5-30/tonne
Treatment	replacement of soils treated	soil washing incineration thermal stripping	£80-120/tonne £400-1200/tonne £40-£350/tonne
	reduce contaminant sources	pumping & treatment of groundwater biological treatment	varies with method £16-80/tonne
		chemical treatment vapour extraction	£135-350/tonne £25-75/tonne
		vitrification cement treatment 'lasagne' multiple-layering treatment	£250-275/tonne £50-100/tonne [not yet viable in the UK]

Croda Hydrocarbons Limited

History

Tar distillation activities have been carried out on the Croda Hydrocarbons Knottingley site since the 1870s, feedstocks being drawn primarily from gasworks and coking plants.

In order to store, segregate and blend tars, eight 'tar wells' were constructed over the years, each being a hole in the ground approximately 12 feet deep and ranging in diameter from 50 feet to 100 feet. The wells were brick lined, surrounded by clay and, originally, provided with wooden roofs.

Following the rapid decline in demand for coal tar due to the advent of North Sea gas and the decline in the steel industry demand for coke, by 1990, coal tar distillation on site ceased and the tar wells became redundant.

Remediation

In 1992 Croda applied to the Department of the Environment for a Derelict Land Reclamation Grant to help fund the cost of emptying the wells and reinstating the ground.

The scope of the resulting grant covered the seven remaining tar wells (one was filled in the 1980s), six of which were visible, but one was underground.

The company awarded the contract for the work to a specialist contractor, whose prime advantage over other contenders was that it proposed an element of recycling in the process. It also had a proven track record of reclamation projects carried out on gasworks or similar tar lagoons.

Liquid material - a mixture of tar, water, oil and emulsions - was pumped into road tankers and removed off site for treatment. The mixture was separated in three phases - water, organics and residues. The residues were landfilled, the aqueous phase treated and discharged to river, and the dry oil phase segregated for re-sale, primarily as fuel.

Work commenced in August 1992 simultaneously on several of the wells.

The first phase of the operation involved the removal of water from the surface of the tar. This was stored in a tank and bled slowly through the on-site effluent treatment system.

The second phase of the job required the use of high capacity pumps to transfer oils and some residues from the wells into road tankers for transportation to the contractor's premises for separation.

Following removal of all material capable of being pumped, amounting to almost 3,000 tonnes in total, pulverised fly ash (PFA) was brought in and machines used to mix residues and ash to produce a solid material with the consistency of tarmacadam suitable for landfill. Total quantity of material ultimately landfilled (residues plus PFA) was over 9,000 tonnes.

Total quantity of stone used to fill the well was over 13,000 tonnes.

The complete contract occupied 27 weeks in total, finishing at the end of February 1993. The total cost of the project was just over £860,000 of which around £600,000 was covered by a Derelict Land Grant.

After a period of settlement, the reinstated ground, mostly in central areas of the site, will be used to accommodate warehousing

Conclusion

Completion of the project in the forecast timescale resulted in a significant improvement in the appearance of the Knottingley site.

Valuable areas of land within the site were released for future development and current operations were facilitated by better access to several production areas.

The removal from the site of all the historical coal tar products has removed a significant potential source of groundwater contamination.

with a consultant with environmental expertise, it is sensible to seek recommendations from lawyers, banks, surveyors and other professional advisers. Note whether the recommendation is of a company or a key individual, who may have moved on. Further help is available from professional associations and directories (see 'Helpful information').

You may like to consider the following criteria:

- Experience of contaminated land assessment and/or remediation for industrial sectors (not just with Local Authorities), including the nature of the specialists on the staff
- Membership of trade associations
- Accreditation of laboratory used (eg. National Measurement Accreditation Service)
- Ability to advise or conduct remedial work (if required)
- Adequacy of professional indemnity insurance
- Terms and conditions of contract
- Price

You may also wish to consider whether to accept a complete package from site investigation to remediation, or to contract stage by stage.

Management

Having chosen the consultants it is important to develop a good working relationship. Set clear objectives. Discuss the work with the consultant to clarify what the investigation will and will not achieve and what effect time and cost constraints will have. A lawyer or surveyor with experience in these matters could assist the process.

Maintain a regular dialogue. In particular, the risk assessment report should be carefully discussed to ensure that any recommended remedial work is justified in terms of necessity and cost. It may be helpful to discuss the draft results of the risk assessment and a final version produced, prior to investigation and reporting of remedial options.

Help with financing remediation

The Government supports the remediation of contaminated land, particularly where it encourages the development of brownfield sites and/or acts as a spur to local economies.

If you think you may be eligible for support for work on your own site, or as part of a partnership, the first step is to consult your local Government Office for the Region. The source and nature of any support which might be available varies greatly with location. The Government Offices for the Regions are responsible for the regional administration of the Single Regeneration Budget which provides around £1.4 billion of public funds for various regeneration projects. Included within the Single Regeneration Budget are:

- English Partnerships, a national agency
- Urban development corporations
- City Challenge

European structural funds may also be available.

English Partnerships has taken responsibility for The Department of the Environment's land reclamation programme. In Wales, the Welsh Development Agency administers land reclamation whilst in Scotland, Scottish Enterprise and Highland and Island Enterprise undertake their own reclamation projects with the involvement of other organisations and their partner Local Enterprise companies.

Contact addresses for the Government offices for the Region, English Partnerships, Welsh Development Agency and Scottish Enterprise are listed in 'Helpful information'.

In 1989 Pilkington, Milverny Properties and St Helens Metropolitan Borough Council formed a partnership to create Greenbank (St Helens) Ltd, to provide a rational basis for treating and reclaiming derelict land south of St. Helens town centre.

Greenbank's initial plan was to reclaim 48 acres of land and former factory works, and to replace them with over 300 homes and 12 acres of new parkland. It approached both the public and private sector to help fund the project.

In October 1989 the Department of the Environment awarded the Greenbank project a £6.3 million City Grant, the largest awarded by Government up to then. This generated a minimum of £24.5 million of new private sector investment.

The reclamation stage started in December 1989 and required the handling of 10 million tonnes of material, including the treatments of mineshafts and the removal of some 300,000 tonnes of waste. Greenbank entered into agreement at the start of the project to sell the land and as a result some funds were granted early to assist funding the reclamation. It also negotiated the contract finance to coincide with receipt of the grants.

Now reclaimed, 100 houses have been sold with a further 280 under phased construction. A new park has been created as an essential element of the environmental upgrading.

The scheme is making a significant contribution to dealing with the problems of derelict and difficult inner city sites in a self-contained and economic way. The linking of private and public sector interests has worked well (at a ratio of 4:1), and Greenbank believes that it can be repeated elsewhere.

Helpful information

Main types of environmental liability

Liability for environment-related damage may arise in many forms and this outline does no more than identify some of the main headings. It is not intended to be a comprehensive statement of the law in this area. If you think that there is any likelihood that you may be liable for damage you should seek legal advice on your specific situation.

[This summary deals only with English law. Scots law is broadly similar but there are some differences. For example, the strict liability principle established in Rylands v. Fletcher does not apply in Scotland and there is not the same distinction between public and private nuisance. Statutory nuisance is dealt with under the Public Health (Scotland) Act 1897. The Water Resources Act also does not apply, with water pollution continuing to be governed by the Control of Pollution Act 1974.]

Liability under statute

Regulatory obligations and civil liability

The main powers and duties of the regulators are found in the Environmental Protection Act 1990 (EPA) and the Water Resources Act 1991 (WRA). With the passage of new legislation creating a single Environment Agency, the statutory powers and duties may be consolidated and updated.

The powers include:

- WRA s.161 - the right of the NRA to recover its reasonable expenses for carrying out work to clean up pollution which has entered, or is likely to enter, controlled waters. The person who caused or knowingly permitted the relevant state of affairs to arise is liable.

- EPA s.79-81 - the right of a local authority (Borough and District Councils) to recover its reasonable expenses in abating a statutory nuisance in respect of which an abatement notice has not been complied with. The person responsible for the nuisance is liable or, where that person cannot be found, the owner or occupier of the property.

- EPA s.27 - HM Inspectorate of Pollution is responsible for Integrated Pollution Control and has powers to remedy harm, enforceable against a person convicted of offences under s.23 of the Act.

The EPA s.61 gives Waste Regulatory Authorities the right to recover from the landowner reasonable expenses for work in relation to damage from deposits of controlled wastes. However, this power is not in force and may be superseded by the powers of the Environment Agency.

Civil liability may arise from the breach of statutory duties, but is enforced by actions brought by private persons in the civil courts rather than by criminal prosecution. Civil liability does not always arise - the Act may make specific mention of the issue or the absence of some other sanction will argue strong in favour of recognising civil liability.

Criminal liability

There is a wide range of primary and secondary legislation governing environmental liability, and much of it is activity, or sector, specific. Under each of the regimes, the offender who commits a criminal offence is generally rendered liable to a maximum £20,000 fine in the Magistrates' Court and to an unlimited sum and/or imprisonment for not more than 2 years in a Crown Court.

Some legislation makes provision for the personal liability of directors, in addition to the liability of their companies, in certain circumstances (s.157 of EPA).

Liability under common law

Liability can arise under four main areas:

- **Private nuisance:** arises where an activity on the defendant's land represents an unreasonable interference with the plaintiff's use or enjoyment of the plaintiff's own land. The plaintiff must prove either physical damage or personal discomfort. Proof of fault appears to be a necessary component of liability, except in the case of interference with natural rights.

  Defences include: right acquired by prescription, right acquired by grant and statutory authority.

- **Public nuisance:** although it is a common law liability it is primarily a crime prosecuted by the Attorney General or Local Authority on behalf of the public. It can be used in private tort action if the plaintiff has suffered particular damage. Unlike private nuisance, damages can be awarded in respect of pure economic loss.

- **Rylands v Fletcher:** This landmark case in 1866 created the situation where someone who accumulates a substance which is likely to cause mischief if it escapes is strictly liable for any damage which it does cause by escaping. The liability only applies where the use of the land is "non-natural". This term had been given an increasingly restrictive definition by the courts, but this trend appears to have been reveresed in the light of the Cambridge Water Company case (see below)

- **Negligence:** A defendant owes a "duty of care" to anyone who is likely to be affected by his actions, and is liable for damage caused by breach of that duty. The damage must have been of a kind that was reasonably foreseeable, and the defendant must have been at fault, although the burden of disproving fault lies with the defendant if damage has been caused by something which would not cause damage if subject to proper care.

Recent developments

Common law

In December 1993 the House of Lords delivered its judgment in the case of Cambridge Water Company vs. Eastern Counties Leather (ECL). Overturning a ruling by the Court of Appeal, it found ECL not liable for damage caused to groundwater, on the ground that the company could not have foreseen that its actions would give rise to damage of the relevant type. However, the terms of the judgment suggested that industrial activity of the type being carried out should no longer be regarded as a 'natural' use of land, thereby potentially having the effect of widening the scope of strict liability under Rylands v Fletcher. The judgment may also have the effect of diminishing the significant doctrinal split between liability in nuisance and under the Rylands v Fletcher principle.

Statute law

The creation of the Environment Agency will require at least a consolidation of some of the main powers, and Government may use the opportunity to adapt them in the light of consultation following the withdrawal of proposals (under s.143 of the EPA) to create a register of land subject to potentially contaminative uses. Subsequently Government can, subject to Parliament, change the powers again as opportunities for legislation arise.

Some new international obligations may also be introduced. In particular, the European Commission published a Green Paper on remedying environmental damage which it plans to develop into a new framework directive on liability.

Funding contacts

Government Offices for the Regions

Eastern
Heron House
49-53 Goldington Road
Bedford
MK40 3LL
Tel: 0234 276106

East Midlands
Cranbrook House
Cranbrook Street
Nottingham
NG1 1EY
Tel: 0602 352342

London
Millbank Tower
21/24 Millbank
London
SW1P 4QU
Tel: 071 217 4657

Merseyside
Graeme House
Derby Square
Liverpool
L2 7SU
Tel: 051 227 4111

North East
Stangate House
Groat Market
Newcastle Upon Tyne
Tel: 091 235 7717

North-West
Sunley Tower
Piccadilly Plaza
Manchester
M1 4BE
Tel: 061 832 9111

South-East
Charles House
375 Kensington High Street
London
W14 8QH
Tel: 071 605 9021

South-West
Tollgate House
Houlton Street
Bristol
BS2 9DJ
Tel: 0272 878126

West Midlands
Five Ways Tower
Frederick Road
Birmingham
B15 1SJ
Tel: 021 626 2733

Yorkshire and Humberside
City House
New Station Street
Leeds
LS1 4JD
Tel: 0532 836608

Other contacts

English Partnerships
16-18 Old Queen Street
London
SW1H 9HP
Tel: 071 976 7070

Welsh Development Agency
Pearl Assurance House
Greyfriars Road
Cardiff
CFl 3XX
Tel: 0222 222666

Scottish Enterprise
120 Bothwell Street
Glasgow
G2 7JP
Tel: 041 248 2700

Highlands & Islands Enterprise
Bridge House
20 Bridge Street
Inverness
IV1 lQR
Tel: 0463 244381

Northern Ireland
Industrial Research and Technology Unit
Industrial Science Centre
14 Antrim Road
Lisburn
BT28 3AL
Tel: 0800 262227

Further contacts

Useful addresses

The British Library
Science Reference and Information Service
25 Southampton Buildings
London
WC2A lAW
Tel: 071 323 7454

Confederation of British Industry
Environmental Affairs Directorate
103 New Oxford Street
London
WClA lDU
Tel: 071 379 7400

Construction Industry Research and Information Association
6 Storey's Gate
Westminster
London SW1P 3AU
Tel: 071 222 8891

Department of the Environment
Contaminated land enquiries
Room A323 Romney House
43 Marsham Street
London
SW1P 3PY
Tel: 071 276 8751

Department of the Environment
Liabilities enquiries
Room A315A Romney House
43 Marsham Street
London
SW1P 3PY
Tel: 071 276 8469

Health and Safety Executive
Baynards House
Chepstow Place
London
W2 4TF
Tel: 071 243 6000

Her Majesty's Inspectorate of Pollution
Romney House
43 Marsham Street
London
SWlP 3PY
Tel: 071 276 8061

National Rivers Authority
Rivers House
Waterside Drive
Aztec West
Almondsbury
Bristol
BS12 4UD
Tel: 0454 624400

Ordnance Survey
Romsey Road
Maybush
Southampton
SO16 4GU
Tel: 0703 792000

Royal Institute of Chartered Surveyors
12 Great George Street
Parliament Square
London
SW1P 3AE
Tel: 071 222 7000

Useful publications

Guidance on 'Why deal with contamination?'

CIRIA (due 1994) *Guidance on the sale and transfer of land which may be affected by contamination*

NRA (1994) *Contaminated land and the water environment*

Royal Institution on Chartered Surveyors (1994) *Manual of Valuation Guidance Notes*

Guidance on 'How to deal with contamination'

Chemical Industries Association (1993) *Contaminated land and land remediation - guidance on the issues and techniques*

CIRIA (due 1994) *The Remedial treatment of Contaminated Land* (12 volumes)

DoE ICRCL 59/83 (second edition July 1987) *Guidance on the Assessment and Redevelopment of Contaminated Land*

Health and Safety Executive (1991) *Protection of Workers and the General Public During the Development of Contaminated Land* (HSG66)

Institute of Petroleum (1993) *Code of Practice for the Investigation and Mitigation of Possible Petroleum-Based Land Contamination*

Institution of Environmental Health Officers (1990) *Development of Contaminated Land - Professional Guidance*

NRA (1994) *Pollution Prevention Guidelines*

Royal Institution of Chartered Surveyors (due 1994) *Guidance on Land Quality Statements*

Welsh Development Agency (1994) *Manual on the Remediation of Contaminated Land*

A bibliography of guidance by the DoE and other regulatory bodies is available from the Association of Environmental Consultancies.

Directories of Environmental consultants

The latest directories, databases and handbooks can be obtained from:

Association of Environmental Consultancies
2 Manchuria Road
London
SW11 6AE
Tel: 071 978 4347

Environment Business
Information for Industry Ltd
521 Old York Road
London
SW18 lTG
Tel: 081 877 9130

Environmental News Data Services
Unit 24 Finsbury Business Centre
40 Bowling Green Lane
London ECIR 0NE
Tel: 071 278 4745

Institute of Environmental Assessment
Gregory Croft House
Fen Road
East Kirkby
Lincolnshire
PE23 4DB
Tel: 0790 763613

ICRCL trigger values for some common contaminants

Note: The values in this table should not be used without reference to conditions and footnotes in ICRCL guidance note 59/83 (second edition July 1987).

Contaminant	**Planned Use**	Trigger Concentration (mg/kg air dried soil)	
		Threshold	**Action**
Inorganic contaminats			
Contaminants Hazardous to Health			
Arsenic	Domestic gardens, allotments	10	n/s
	Parks, playing fields, open spaces	40	n/s
Cadmium	Domestic gardens, allotments	3	n/s
	Parks, playing fields, open spaces	15	n/s
Chromium	Domestic gardens, allotments	25	n/s
(hexavalent)	Parks, playing fields, open spaces		
Chromium	Domestic gardens, allotments	600	n/s
(total)	Parks, playing fields, open spaces	1000	n/s
Lead	Domestic gardens, allotments	500	n/s
	Parks, playing fields, open spaces	2000	n/s
Mercury	Domestic gardens allotments	1	n/s
	Parks, playing fields, open spaces	20	n/s
Selenium	Domestic gardens, allotments	3	n/s
	Parks, playing fields, open spaces	6	n/s
Phytotoxic contaminants			
Boron (water soluble)	Anywhere plants are to be grown	3	n/s
Copper	Anywhere plants are to be grown	130	n/s
Nickel	Anywhere plants are to be grown	70	n/s
Zinc	Anywhere plants are to be grown	300	n/s
Organic contaminants			
Polyaromatic	Domestic gardens, allotments, play areas	50	500
Hydrocarbons	Landscaped areas, buildings, hard cover	1000	10000
Phenols	Domestic gardens, allotments	5	200
	Landscaped areas, buildings, hard cover	5	1000
Free cyanide	Domestic gardens allotments landscaped areas	25	500
	Buildings, hard cover	100	500
Complex	Domestic gardens, allotments	250	1000
cyanides	Landscaped areas	250	5000
	Buildings, hard cover	250	n/l
Thiocyanate	All proposed uses	50	n/l

Sulphate	Domestic gardens, allotments landscaped areas	2000	10000
	Buildings	2000	50000
	Hard cover	2000	n/l
Sulphide	All proposed uses	250	1000
Sulphur	All proposed uses	5000	20000
Acidity	Domestic gardens, allotments, Landscaped areas	pH<5	ph<3
	Buildings, hard cover	nil	n/l

n/s: not specified

n/l: no limit set as contaminant is not particular hazard for this use